FLORA OF TROPICAL EAST AFRICA

BUTOMACEAE

Susan Carter

Perennial or annual, aquatic, swamp or marsh herbs, glabrous, usually lactiferous. Rhizome short ; roots short, fibrous. Leaves erect or floating, basal ; petiole with a sheathing base ; leaf-blade entire, iridaceous, or lanceolate to orbicular with cuneate to truncate base and acute to rounded apex. Inflorescence umbellate, rarely with solitary flowers ; bracts 2 or 3 ; bracteoles several ; flowers regular, bisexual. Sepals 3, persistent. Petals 3, delicate. Stamens 6–9–∞ ; filaments flattened ; anthers 2-celled, dehiscing longitudinally and laterally. Carpels superior, free or joined at the base, 6–∞, in a whorl, unilocular ; style terminal ; stigma sessile ; ovules ∞, scattered over the ovary-wall on a reticulate placenta. Fruiting carpels finally dehiscing along the ventral suture ; seeds ∞, smooth, wrinkled or ridged, rarely slightly spiny, without endosperm ; embryo horseshoe-shaped or straight.

A small family, confined to the tropics or subtropics except for *Butomus* L. which extends across Europe and Asia. The following genus is the only one which occurs in Africa.

TENAGOCHARIS

Hochst. in Flora 24 : 369 (28 June 1841)* ; F.W.T.A. 2 : 298 (1936)

Butomopsis Kunth, Enum. Pl. 3 : 164 (July 1841) ; F.T.A. 8 : 214 (1902)

Perennial marsh herbs. Leaves erect. Leaf-blade oblanceolate, slightly acuminate. Peduncle longer than the leaves ; inflorescence umbellate, of 1 whorl of flowers, sometimes 2 in larger plants ; 2 bracts and several bracteoles at the base of each whorl. Sepals 3, broadly ovate, erect. Petals 3, larger than the sepals, withering and disintegrating after anthesis. Stamens about 9 ; filaments flattened below ; anthers oblong, narrow. Carpels 5–8, joined at the base. Fruiting carpels exceeding the sepals ; seeds many, black to brown, smooth and shiny ; embryo horseshoe-shaped.

Monospecific, widespread in the tropics and subtropics, but absent from the New World.

T. latifolia (*D. Don*) *Buchen.* in Abh. Nat. Ver. Bremen 2 : 1 (1868) ; E.P. IV, 16 : 6 (1903) ; F.W.T.A. 2 : 298, fig. 278 (1936). Type : Nepal, Hamilton‡ (location unknown)

Petioles (5–)10–20(–25) cm. long ; leaf-blade 3–11(–15) × 1·5–3·5 (–5) cm. ; base cuneate ; apex bluntly acuminate ; nerves 3–7. Whorls of 3–11 flowers, rarely up to 20 ; pedicels 6–11 cm. long, angular ; bracts membranous, triangular-lanceolate, up to 1·5 × 0·5 cm. ; bracteoles about 6, membranous, smaller than the bracts. Sepals up to 7 × 4 mm. Petals white, very delicate. Stamens with filaments 2 mm. long elongating to 4 mm. long at maturity ; anthers 2 mm. long. Carpels about 5 × 2 mm., with

1

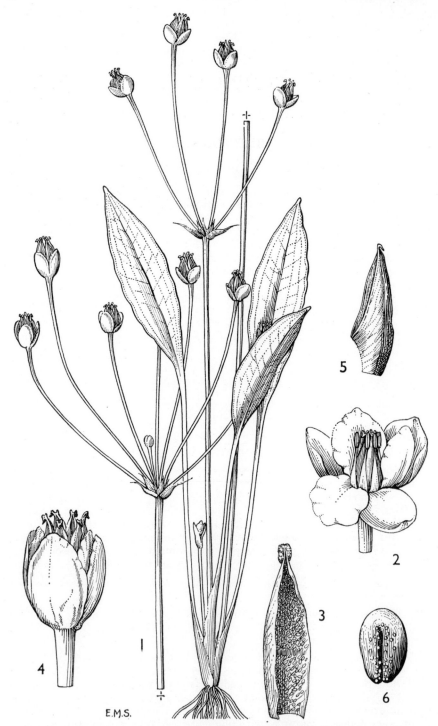

E.M.S.

FIG. 1. *TENAGOCHARIS LATIFOLIA*—1, plant in fruit, × 1 ; 2, flower with a sepal and petal pulled down, × 3 ; 3, young carpel opened out, × 9 ; 4, fruit, × 3 ; 5, ripe carpel after dehiscence, × 3 ; 6, seed, × 60. 1, from *Lea* 220 ; 2–4, from *H. B. Johnston* IV. 82 ; 5, 6, from *Lind* 260.

sessile stigmas. Fruiting carpels 9–12 mm. long. Seeds less than 0·5 mm. long, minutely warted. Fig. 1.

UGANDA. Teso District : Soroti, 14 Sept. 1954, *Lind* 260 !
DISTR. **U3** ; known in East Africa from one gathering only, but occurs from Senegal and Ghana to the Sudan ; also in India and very rare in the Malay Islands and N. Australia.
HAB. Marshes and swamps ; 1080 m.

SYN. *Butomus latifolius* D. Don, Fl. Nepal 22 (1825)
    B. lanceolatus Roxb., Fl. Ind. 2 : 315 (1832). Type : Bengal, *Buchanan‡ in Wallich* 4999B (K–W, ? syn. !)
    Tenagocharis cordofana Hochst. in Flora 24 : 369 (1841). Type : Sudan, Kordofan, Arashkol, *Kotschy* 193 (BM, K, iso. !)
    Butomopsis latifolia (D. Don) Kunth, Enum. Pl. 3 : 165 (1841)
    B. lanceolata (Roxb.) Kunth, Enum. Pl. 3 : 165 (1841)
    B. cordofana (Hochst.) Kunth in Walp., Ann. 1 : 769 (1849)

*See W. T. Stearn in Fl. Malesiana, ser. 1, 4 : CXCV (1954).
‡Francis Buchanan (1762–1829) assumed the name of Hamilton in 1818. His name is cited as " Buch.-Ham." when used as an authority for a botanical name.

INDEX TO BUTOMACEAE